Prolegomenon Towards A Primate Rights Bill

Freedom 4 Animals

EPIS Press
2026 S. 9th W. #4
Missoula, MT 59801 USA
animalsRfree@gmail.com

Freedom 4 Animals is an imprint of EPIS Press

Freedom 4A nimals name and logo
are trademarks of EPIS Press

Printed in the United States of America
First Edition

Library of Congress Cataloging-in-Publication Data
1. Animal Rights 2. Bill of Rights 3. Animal Research models
4. Humanism 5. Speciesism 6. Primates I. Title

Cover Design: Tia Hopkins
Author Photo: NTG
Author Seal: Adrian Balasa
Freedom 4 Animals logo: Francene Higman
ISBN 978-1-943332-01-4

Prolegomenon Toward A Primate Rights Bill

By Dr. Kevin Boileau
&
Nazarita Goldhammer

Dedication

We dedicate this prolegomenon to all the primates that human beings have savagely harmed, abused, neglected, and forced into servitude or imprisonment, both past and present. We hope that wiser humans will follow and support this prolegomenon with an actual primate rights bill that would become federal law in the United States, and that would influence similar laws around the planet Earth.

Introduction

It is recognized that all primates, including human beings, have a common evolutionary origin even though each primate group has differentiated itself from the others, and even though the human group has differentiated to the greatest extent.

It is recognized that all primates have similar nervous systems and capacities for autonomy and suffering.

It is recognized that all primates have complex sets of interests, needs, and preferences, and that the recognition of natural rights protects these. Therefore, all primates and all primate groups have natural rights.

It is recognized that the very existence of a primate group implies that it has the natural right to continue to live free from interference from this living by human primate groups.

It is recognized that the ignorance of disrespect for these natural rights causes serious harm to primate individuals and groups; to nature as a whole; and to human beings.

It is recognized that the respect for the natural rights of all primate groups is inextricably tied to the respect for the natural rights of all human primates.

Preamble

Part 1

This prolegomenon applies to all beings who are in the class of primates.

We recognize that there has been a historical evolution in the conceptual and legal distinction between personhood and property. For example, we recognize people of certain race, color, age, who have long suffered invidious discrimination, including African Americans, women, children, the disabled, laboratory fetuses, and we recognize that non-human sentient animals such as primates have also been affected and harmed by this distinction.

For example, in some states, dogs are viewed as cash crops, treated no differently from oranges. We also recognize there is an analysis on the moral plane, in which non-human sentient animals have the same relevant, moral characteristics as human beings, but who have also been treated dissimilarly. For example, the class of beings called primates, which includes but which is not limited to orangutans, bonobos, monkeys, chimpanzees, who are individually autonomous, who are capable of suffering and who are aware of their suffering, are treated differently from human primates. There is a historical and analytic component to this understanding of a system of morality, which includes dimensions of the social, political and scientific that treats similar cases dissimilarly. This is the very definition of injustice.

This prolegomenon focuses on this injustice and this analytically illogical treatment and historically illogical treatment of non-human, sentient primates. Its various provisions both separate and collective, intends to correct this injustice by promulgating proto-federal law that protects the rights and the interests of all primates to be free from physical and psychological harm and, moreover, mandates an actual responsibility and requirement that human beings acknowledge, promote, and serve the best interest of all primates in the world today whether they are living in the wild, in a research facility, in any other government organization, in a private home, or in any other place.

It is important to point out that these natural and legal rights of primates do not depend on whether they are considered to be persons or not. The status of persons is based on a Kantian moral system in which rights and duties are considered to exist together, correlatively. This means that for a person to have rights he must also have duties. Given that primates do not have duties it is supposed that they cannot have rights. Therefore, they cannot be persons. However, in the bill that we promulgate here, the natural and legal rights of primates do not depend upon this system of correlative rights and duties. Instead, the foundation is that natural rights come from the faculty of autonomy, sentience, and the potential to suffer. This does not depend upon the assignment of the status of personhood.

It should be absolutely clear that we recognize both natural and legal rights of all primates, including but not limited to human beings because there are no morally relevant differences between one type of primate and any other type of primate given that they all are capable of autonomy, sentience and suffering.

Human rights declarations focus only on the natural rights of human beings, but they ignore the natural rights of other primates to be free from harm and to pursue a natural life pattern as part of a vital role in all eco-systems; to enjoy the common faculty of autonomy, choose life plans, and seek shelter, comfort, and adequate food; and to live amongst their own species in nature family groups. From these natural rights, which are no different from the natural rights of human beings—save the desire to form nation states and human laws—we hereby promulgate the following legal precepts that will protect, serve and foster the natural rights we hereby acknowledge. It is our goal to produce several successive iterations of this proto-law, with the intent that its principles eventually be promulgated as United States federal law.

Preamble

Part 2

If we compare the *Universal Declaration of Human Rights* to the *Universal Declaration of Animal Rights* we can see that they make assumptions that the beings to be protected have natural rights. According to the documents, these rights come from having certain characteristics and then are promulgated by fiat. The methods are exactly the same: certain characteristics lead to natural rights. In the case of human rights, non-human animals are ignored. In the case of animal rights, humans are ignored.

We can see, however, that there is no relevant difference between human primates and other sentient beings who are classified as primates. Logically, this requires us, therefore, to include all sentient beings within an account of rights even though we have not yet done so historically. The Universal Declaration of Human Rights, the "UDHR," calls attention to certain features of human beings, such as "inherent dignity," "equal and inalienable rights," a "right to life, liberty, and security," a right not to be "subjected to torture or to cruel or degrading treatment or punishment," and a "right to freedom of movement." These speak of a high valuation of human beings that comes just from their identity as humans. It comes from our natural conceptions about human life and our recognition that we must protect important values and goods. In the case of humans, we simply agree that human life is valuable, attaching various positive terms to this valuation. It is a pronouncement by fiat.

In this process we ignore other sentient life, tacitly agreeing that other life is not important enough to include in various statements about natural rights. Historically, animals have not been thought to deserve inherent rights, only gaining value in terms of their instrumental value to humans (which, by the way, Immanuel Kant acknowledges). Yet this does not mean that animals do not have inherent dignity just like humans, or any of the other protections that are listed above. Let's consider what could account for this.

Historically, we understand that the kind of self that we have created valorizes the development of the autonomous self in which each of us lives in the world in terms of his own mental categories. In this tradition we approach Others in terms of our own subjective interpretation of meaning and analyze Others relative to what we already know. It is an appetitive approach in which we consume Others by reducing them to the same cognitive categories of our own minds and experience.

This is an anthropological mistake that operates both between humans and also between humans and other sentient beings. More specifically, it operates between humans and other primates. We have seen how this anthropological mistake operates in an intra-human context. Now, let's explain how it works when we consider other sentient beings. First, in a humanism of the "I," this masterful, egocentric self tries to dominate everything in its path, which closes the natural capaciousness of consciousness into a direction that ends in destruction and horror. This is the consciousness that callously disregards the interests of all animals, most humans, and many primates. You see, just because we come to idolize the ideals of humanism we fall into a self-deception about the praxis. Upon inspection, we see that humans don't actually do very well in meeting the demands set forth in a universal declaration of human rights because we don't often protect the inherent and natural dignity of other humans. This is the state of consciousness that is selfish, egocentric, violent, and narcissistic. Yet, unfortunately, the discussion stops at the borders of the human and all consideration for other sentient beings is lost. It appears very clear that the structure of consciousness in egocentric narcissism by humans toward other humans, is exactly the same structure in behavior by humans toward animals. It is the very same egocentrism that causes a human to act violently toward another human that also causes him to preclude animals from a statement of natural rights and therefore, to give implied permission for their instrumental use and violent death. This is the psychological structure that we overcome in an interpretation of our human anthropology.

A secondary argument that accomplishes the same goal involves rationality. The argument has often been made that because animals cannot reason they are of lesser ontological worth than humans. However, this very type of consciousness bleeds back into intra-human affairs in which we subtly come to believe that the more "rational" one is the more he or she is worth. Thus, humans with severe birth defects in which their cognitive abilities are seriously impaired are implicitly viewed as having less value than those with normal or superior abilities (especially in a utilitarian moral system). The defective human (and this extends to those in prison, those who have menial jobs and the like) and the animal rate lower in value than those humans who have better rational capacities. It is a specious argument, and its consequences only serve to fuel the perennial "war of all against all" in our civilization. This is Thrasymachus's position in the *Republic,* which Plato shows all too easily cannot serve as a foundation for justice. Moreover, in a derivative way, we can also see how this very same structure of consciousness is foundation to the neglect, abuse, and destruction of the environment. A trans-humanist foundation that treats like cases in like manner amongst the full spectrum of primates challenges this way of thinking.

For purposes of this bill, we will focus on humans and other sentient beings, and leave the environment in general to another day. We shall also leave to another day the definitions of sentient being, animal, and those creatures who/that do not achieve such a denomination. As far as the natural rights of humans go, we engage in a great deal of deception. It seems clear that the notion of [equal] human rights becomes more of something to say—to use in a manipulative way—than to actually practice. Perhaps this stems from a true lack of the ontological parity thesis, the result of which leads to violence toward any Other. In short, there is no basic ontological difference between humans and any other feeling animal that would lead to different moral treatment. We also believe that the exclusion of animals from a bill of natural rights comes from the same thinking that creates a bill of rights for humans but that does not actually respect it. There is also a third argument that does not have any convincing effect whatsoever. This is the argument that in order to have rights one must have duties, for this is what creates a reciprocal system. It is, instead, the case that there are many people who have rights but who are not capable of exercising duties. These people are the elderly and the mentally or physically infirm. Thus, it makes no sense to distinguish rights on this basis, and to conclude that because non-human animals cannot enter into the rights-duties system they don't have rights. In effect, this continues the "might makes right" mentality and does not get us to a better position than the very narcissistic, egocentric state of mental development. Let us, therefore, leave this dimension and move to the phenomenological aspect that we can articulate through a discourse about solidarity. As a bridge to this discussion, however, we offer an idea about a new foundation for rights for both humans and animals that emerge from a common root. This is the root of life. This is vivantonomy, a new concept that expresses a new human anthropology and a new understanding of the place of humans on the planet Earth. This is a reconstituted humanism.

In the Preamble to the *Universal Declaration of Animal Rights* it is stated "that Life is one, all living beings having a common origin and having diversified in the course of the evolution of the species"; "that all living beings possess natural rights"; [and further in the Preamble] "that the respect of humans for animals is inseparable from the respect of man for another man." The declaration goes on to specifically delineate 14 articles that address the notion that animal rights and human rights come from the more primordial dimension of life itself. We have seen how humans have developed a declaration of human rights from the notion of natural rights, although animals and the environment are excluded. In this declaration for animal rights, we recognize a great diversity in living creatures, and specifically acknowledge that each of one has natural rights. These rights come from the sole fact of being alive, and participating in a wondrous, complex eco-system if you will, but more than that, representing a different facet of the interrelatedness of all living beings. We can see that a declaration of human rights that bases its strictures on notions of natural rights cannot escape the fact that natural rights also belong to other sentient beings. Our problem is that we have either neatly ignored or forgotten this fact. or we have callously disregarded it (both a product of the ego-centric mind). We have prepared the reader to overcome these limitations through the provisions of the bill.

Preamble

Part 3

Preamble

Part 3

There are many species of beings that we humans will never see, and many that are in our daily ecosystems that we choose not to see. Yet, they are there—here—rather, constantly watching, looking, appealing—usually to us humans. We don't see them because our own systems of value are informed by our narcissism: our egocentrism. This type of consciousness therefore closes itself in on itself, not seeing other life, other humans, and our very selves. This is the possessory, dominating subjectivity that instrumentalizes all others, and even in a system indoctrinated by rights and duties, fails to see the Other's world on its own terms, as its unique manifestation. Here are the 6 axioms that underwrite an understanding of the principles of the natural and legal rights of primates:

First, we must dismiss the notion that humans are better, of more worth, or higher on a value scale. We must substitute it with a new axiom of ontological parity.

Second, we must agree that in principle that most of us have little knowledge about the whole: about how all beings, processes, and structures work together in an ecosystem. We substitute it with a new axiom of rigorous inquiry.

Third, we must accept a new Archimedean point. We cannot pretend to be at the center of the universe or the planet earth. This means that we must render an accounting of all life forms, including ours, holding that all living beings have equal interests and rights. We must, therefore, have an axiom that recognizes we play a part in the whole but are not the whole, and that we must mediate and weigh our interests relative to those of other life forms. If we do conclude that we are the center of the planet then we must reject our species narcissism and our hubris, and replace it with a strong sense of responsibility to all other life forms and to all other primates and primate groups.

Fourth, we must recognize that all life forms come from the same source. This leads us to the reconstituted notion of solidarity. This is a trans-human notion that includes all other life forms equally with the human.

Fifth, we must acknowledge and accept a new depth and breadth of our responsibility to others, including humans, other sentient life forms, additional life forms, and the environment in general.

Sixth, we must work diligently to formulate and articulate a new philosophical anthropology for human beings. This means we must strive for new meaning and understanding of the world and our place within it. This is neither the autonomous subject nor the heteronomous subject but it is a new human. This re-formulates the reality principle. Here are the 31 articles of this Prolegomenon, followed by a philosophical exegesis of their foundation.

The Principles

Article I

All primates are sentient beings who are born free and equal in dignity and rights. They are endowed with reason, autonomy, needs for family groups and social structure, and adequate resources such as water, healthy food, and the ability to move around in accordance with natural rhythms and life ontologies.

Article 2

Every primate is entitled to all the rights and freedoms set forth in this Prolegomena without distinctions of species of any kind. Human beings must respect the natural rights of all primate groups even though the human group might be considered to be superior in some ways. This superiority may not give humans superior rights.

Article 3

Every primate has the right to a healthy, natural life in accordance with its normal and regular functions, rhythms, processes, life cycle, other cycles, and structures; liberty and autonomy; and security of his or her body, themselves, and their family groups. Human primates must respect this right.

Article 4

No primate shall be held in slavery or servitude; slavery and the slave trade shall be prohibited in all their forms. No primate should be treated as a physical object ever. Whether or not a primate is a legal person under the laws of the U.S. is irrelevant. What is relevant is their autonomy and capacity for suffering. Humans should never hold any primate, including other humans, in slavery, for any reason.

Article 5

No primate shall be subjected to physical or psychological torture, punishment, or any other degrading or harmful behavior for any reason, including the reasons of education, understanding, and scientific research.

Article 6

Every primate has the legal right to the public recognition that it has interests that must be protected by law whether or not non-human primates are considered to be "persons" with due process rights in our U.S. and state constitutions.

Article 7

All primates are equal before the law and are entitled without any discrimination to equal protection of the law equal to that of human primates, whether or not some primate groups are not considered to be persons. This is tantamount to a revised bill of rights for all primates not just some of them. The rights of one primate group may not be compromised for the interests of any other primate group.

Article 8

Every primate has the right to an effective remedy by a competent national tribunal for acts violating the fundamental rights granted him by the constitution or by law. This means each primate has standing under the law to bring a court action for a legal remedy whether in person or by proxy. This includes the right to legal representation to redress any harm done to that individual primate.

Article 9

No primate shall be subjected to arbitrary arrest, detention or exile, which includes locating an individual primate in isolation in a lab, dwelling structure, institutional captivity, or otherwise. Each and every primate has rights to due process equivalent with those that human beings enjoy. Each and every primate has rights of equal protection as a suspect class with full judicial review that accords with the standard of strict scrutiny.

Article 10

Every primate is entitled in full equality to a fair and public hearing by an independent and impartial tribunal, in the determination of his rights in any case where his rights might be affected.

Article 11

No primate, except a human primate, may be charged with a penal offense. By law, non-human primates are incapable of the mens rea required for a criminal offense.

Article 12

No primate shall be subjected to arbitrary interference with his privacy, family, or home. Every primate has the right to the protection of the law against such interference.

Article 13

Every primate has the right to freedom of movement and residence within his home/native environment, certified primate sanctuary, or equivalent accommodation that protects the rights laid out in this document.

Article 14

Every primate has the right to seek through human proxies the freedom from persecution and harm in a designated primate sanctuary or other location that affords them all the other freedoms, rights, and enjoyments outlined in this document. The U.S. should seek international support from the United Nations to extend these protections to other countries. The U.S. shall also follow the example set by other countries that currently afford primates more rights and protections.

Article 15

Every primate has the right of individuality, a name, and personal designation as a protected being under the laws of the U.S. No primate shall be deprived of the right of individuality, a name, and personal designation by any state or federal government or individual human primate.

Article 16

Primates have the right to choose mates, to procreate, and to create family groups. It must be recognized that having a family is a natural and fundamental right for any and all primate groups and individuals. It is also recognized that humans must make affirmative steps to protect the natural and legal rights of family groups.

Article 17

All primates have the natural and legal right to possess the various forms of property that they require for their healthy existence and for their flourishing in healthy ways that are consonant with their basic and natural life rhythms and goals. No primate shall be intentional deprived of possessions that are necessary for his or her autonomous existence and flourishing. This includes both real property and personal property.

Article 18

Every primate has the right to freedom of thought either alone or in community with other primates, and to manifest his belief in teaching and practice in accord with the values extant in his culture.

Article 19

Every primate has the right to freedom of expression without interference from any other primate group.

Article 20

All primates have the freedom to associate with others in their primate group or in any other primate group.

Article 21

Each primate has the autonomous right to participate in his family, social and cultural group and its leadership, and to enjoy their full nature, interests, and autonomous decisions within these groups without harmful interference from any other primate group or individual. Humans have the affirmative duty and responsibility to not interfere in these primate groups in harmful ways.

Article 22

Every primate has the right, as a member of his society, to be provided adequate resources for the full development of his individuality and autonomy, and his membership in a family group and social collective.

Article 23

Every primate has the right to work without interference from human primates, for the benefit of himself and/or his family and social group.

Article 24

All primates have the right to rest and leisure, and therefore should never be enslaved to work for the benefit of any other primate group.

Article 25

All primates have a right to a standard of living that allows them their normal functions and natural rhythms, including water, food, living conditions, health care, socials services, and security in the event of sickness, disability, widowhood, old age or other lack of ability or condition. All children and mothers should be afforded specific and focused care if necessary.

Article 26

All primates have the right not to be prevented from experiencing the education they normally receive from elders in their family groups and social groups. It must be acknowledged that this education is naturally directed to the full development of their natural life expectations; their personality; their respect for their family groups and social groups; and their respect for the strengthening of their fundamental freedoms. It should also be respect that primate parents have the natural right to direct the education of their young.

Article 27

All primates are entitled to a national legal structure that supports, protects, and develops these natural and legal rights as said.

Article 28

Human primates have the primary duty of affirmatively acting to secure respect for the rights and freedom of all other primates; to create conceptual systems of morality that consider natural rights of all primates; and to create integrative relationships between all classes of primates.

Article 29

No scientific research should ever be performed on non-human primates unless it directly leads to the best interest and benefit of that individual primate upon which the research is performed. Research that harms one primate group for the benefit of another primate group is unjustified, unacceptable, and prohibited.

Article 30

Human primates should not ever use non-human primates for human entertainment.

Article 31

No individual, group, organization, or government in the U.S. has the right to engage in any activity or to perform any act aimed at the destruction of any of the rights and freedoms set forth herein.

Philosophical Exegesis of Responsibility and Solidarity

Foundation for a Primates Law and Federal Bill

What follows is a philosophical exegesis of the underlying concepts of responsibility and solidarity that support such a primate rights bill. It shall inform individuals who refine our work into federal law of the deep commitment we must make to other and all primates. This foundation is "trans-humanist," which means that it is a reconstitution of our current understanding of our humanity. In effect, it alters the anthropological course of humans.

Solidarity

In order to develop the concept of a trans-humanist philosophical anthropology that extends a radical heteronomous solution to the problem of our behavior toward other humans, the environment, **and non-human primates**, we need to explore the concept of solidarity. In order to clarify how we think of solidarity it will be helpful to consider a new view of human subjectivity that prioritizes the Other, in which the Other becomes the foundation for my freedom. We also prioritize solidarity over other important values such as freedom and equality because we think they depend on it, as first principle. This prioritization transmutes the kind of autonomous individualism of the Enlightenment to the sort of heteronomous subjectivity that this new kind of valuation supports, one that goes beyond human interests. Let us now develop an account of solidarity-fraternity that is based upon these notions. We will explain how solidarity is a complex value and virtue that can found human rights and primate rights.

As foundation to this new approach to the moral life and legal protection for all primates, it is important not only to do the right thing but also to have good dispositions, motivations, attitudes, character, and moral habit. It goes beyond the duties of the law and motivates people toward excellence. The overall goal is for humans to go beyond mere survival and to flourish together in communities that include other sentient life forms such as non-human primates. This sought after state of *eudaimonia* requires social institutions that work for justice and fairness. This infrastructure creates the environmental conditions that allow for the individuals within it to excel.

Once we identify the character traits we ought to cultivate, then we must consistently incorporate them into our personal, professional, social, and political lives. When we acquire these dispositions we must then learn how to apply them to the right degree. That is, in our experience of such states as fear, anxiety, anger, and so forth, we must learn to experience them at the right time, toward the right objects, toward the right people, for the right reason and in the right manner, applying a standard of balance. What is very important is that we emulate those of us who have developed virtuous behavior. Correspondingly, when we act virtuously we must serve as exemplars for others in our communities.

In order to determine which of the virtues we ought to focus on and develop, we must first understand our values as a community, society, or ecosystem, for it is these values that virtuous behavior must serve. In the case of this bill, we focus generally on all beings that comprise an ecosystem, and specifically on primates, who are important members of various ecosystems, and essential beings in the set of all ecosystems, the supra-system of the planet Earth. Let us understand that a value is an ideal or something that we aspire toward as a state of being or a condition. In contrast, a virtue is a behavior that supports the value. A partial list of possible values could include the following ones: unity, identity, democracy, integrity, family, community, hard work, and strength. Let's briefly define them.

Unity means a kind of solidarity in which we come together to advance the dignity of each of us, and all of us together. This does not seem to be the type of system that has emerged from the Enlightenment and its valorization of autonomy. As a part of this solidarity we must acknowledge the identity of each individual and group of individuals, including all primate groups and individuals. To further the value of identity, we must also value democracy, for in this kind of system we respect diversity, our responsibility to it as well as to the democratic process. Integrity implies unwavering commitment. Hard work and strength imply assiduousness, focus, and perseverance. Further, there are many ways to talk about this list of values and their corresponding virtues. We must focus here, though, on how this cluster of values comes together to form what we are as moral selves. While keeping the preceding list in mind, let's focus on three major values as well as related values that support the major ones. The major values that we want to address are solidarity, justice, and responsibility. The supporting values include integrity, respect for others, beneficence, and compassion.

The primordial condition of solidarity is that the other is entrusted to me in an unconditional manner. We are similarly situated, with similar pain, and similar challenges. It is a belonging together that emerges from the realization that the community one belongs to constitutes a single whole. It is a state of interdependence, an ethical feeling of compassion, and a deep spiritual connection. On the sociological level, for example in any community, group, or organization, we are dependent upon each other, not only for our survival but for our flourishing as well. On the personal level this requires us to be empathic or compassionate with each other, to stand in each other's shoes. This requires us to commit ourselves to the common good and to the good of each individual because each of us is responsible for all. Even more, it requires radical selflessness and forgiveness of each other. It presupposes a symmetrical relation of respect for each other's rights, as well as an asymmetrical relation that goes beyond rights. This asymmetry requires self-sacrifice. Thus, the needs of the Other take priority over one's own interests. In other terms, we must take care of the whole, the common interests of all, and the best interests of the Other, always, as exemplification of this solidarity. This common good includes the protection of all primate groups, not just human primates. This is because we comprise the unity of primates, share all much of the same needs and desires, and are all part of the same origin, part of the interdependence of the world.

Solidarity requires us to go beyond the contractual model to a social-behavioral model that is based on covenant. In this model, every relationship that is mediated by objects, including capital and property, must be subordinated to the more fundamental relationship between humans as communal beings—that is, we need each other. However, we must not overlook our deeper relations to other beings outside of the human community. This is at the level of ecosystems that we share with many other beings, and especially non-human primate groups. The quality of the relationship of solidarity is an end in itself, a state in which we overcome the valuation of others for extrinsic, instrumental reasons. This is linked to the axiom that there is a universal right to use the resources of the earth but in a way that respects and honors the use of these resources by other species. Thus, based on the assumption of the individual dignity of each person, each of us must be seen in a way such that individual interests must never run counter to the general welfare. That is, the universal goods of the union are to be shared universally. This requires those of us with more shared goods to share them with the less fortunate. This process should humanize our contractual relations. In a trans-humanist account, we must also take into account the equal dignity of all living beings and, in this case, the equality between all primates. We must limit our use of the earth's resources and the treatment of non-human primates by the interests of all primates, not only by human interests. This means that the values of money and commodification must never take priority over the important life interests of primate groups. In short, humans should never view their own interests as greater than the interests of any other primate group.

The process of contributing to this state of solidarity triggers the value of justice. Justice is fairness. It is the principle of moral rightness. It is the upholding of what is just, especially fair treatment and due reward. In valuing justice, then, for the purposes of this bill, we are committed to the fair and equitable treatment of all primates. Justice entails rewarding human persons in accordance with their deserts, or withholding the reward when appropriate. It also entails punishment for those who deserve it. This system of rewarding and punishing ought to be carried out by those in power in accord with substantive and procedural rules. This raises the issue of whether those in power were elected to positions of power through fair processes and whether the system of reward and punishment is itself fair. Further, justice is not so much about equality as it is about equity. Thus, sometimes people should be treated equally and sometimes not. The three main types are retributive, compensatory, and distributive. Justice as retribution concerns fair punishment practices. Typically, in the discourse of ethics we refer to retribution in terms of accountability and the need to punish those who violate professional standards. Part of this involves exposing those who violate our common standards and agreements. Compensatory justice concerns fair reward practices. It is what we do when we settle on an equitable way of rewarding individuals for their achievements or benefiting others because of their disadvantages or for harms done to them. Finally, distributive justice concerns fair allocation practices. It concerns how the benefits and burdens of our community are passed out. In our community it concerns the allocation of protections, rights, goods, and services. In a trans-humanist account, we must extend this conception of justice to all primate groups, not only to human primates. This is the shift in Archimedean point this bill advocates. **In terms of the three forms of justice, this shift in perspective excludes retributive justice because non-human primates lack the *mens rea* that would require retribution. However, both distributive and compensatory justice ought to encompass and consider all primates, both human and non-human types. To not do this violates these forms of justice.**

When we translate the value of justice into action we usually must make a comparison. We compare the actions, needs, or merits of one person to those of others. In order to assess what is fair we must assess what a person deserves by comparing him with others in some relevant way. We can only determine what is fair by determining what is fair to others. Thus, justice is always relative to the situation, persons and primates, and resources at hand. Justice also requires equal treatment of those who are similarly situated and unequal treatment of those who are not similarly situated. In the attempt to be just we must presume that persons are equals unless reflective differences show that they are not. In order to demonstrate that persons ought to be treated differently, for example with regard to life needs, preferences, and standards of flourishing, we must analyze and compare. As we have stated previously, all primates have the same basic needs. **Thus, the same kind of legal and moral analysis we make amongst and within human groups ought to apply to all primate groups, in accordance with our conceptions of justice, other than retributory justice, which does not apply. At a bare minimum, we must engage in a moral analysis that goes beyond utilitarian or any other kinds of analysis that do not assume ontological parity between all individual members and groups of primates. To do otherwise, means to think and act unjustly, in principle. Let us now discuss responsibility.**

Responsibility

In order to pursue justice, we must overcome a concept of self that limits responsibility, which is the next value that we want to adopt. Responsibility is first that quality that makes us committed to anything. It requires that we follow through on the values that we have recognized as a basic part of what it means to be human and more specifically what it means to be a part of any particular group. When we are connected to other people we have responsibility to them. It is what makes us care about being ethical and perhaps can be spoken about as moral motivation. Responsibility is perhaps the most important value underpinning any moral transformation. Let's therefore develop an extended and sustained discussion about it. In order to prepare for this discussion, we must imagine a social ontology—a set of expectations—that is not merely based on contract. Let's consider instead that it is based on the axiom of ontological parity.

Responsibility can be thought of in two senses. First, it is the precondition for being moral, that is, thinking that ethics is important in the first place. Second, it is the list of those specific actions required of us when we are responsible to or for someone or something. It is that which binds us to the world and to others, requiring a response when appropriate. When we are infants we do not understand responsibility; our only task it seems is to immediately satisfy our desires. As we grow older we recognize that our needs must be rank ordered and mediated by the needs of others. Responsibility manifests itself by reflecting on and acting upon the constellation of our whole value system. It is the very way that we can explain our connection to Others. Further, it is the way that we connect to the world and to Others--in terms of our value systems--that gives meaning to the lives of each of us. Thus, responsibility includes a relational component in which we reflect on the needs of Others; in the case of trans-humanism, we consider the needs of all other primates.

Our specific responsibilities come in two ways. We are responsible to some Others, for example our spouses, friends, family members, and colleagues, and we are responsible for some Others, including members of that same overall group mentioned. When we are responsible to someone we must render an accounting to that person—that sentient being—and we have duties to do so. These duties in fact come from the very nature of the relationship that we have presumably entered into freely. For example, a mailman has the duty to deliver the mail even though he may be tired or not feel well. A mother may have the duty to care for her children even though she may have other interests and desires. These responsibilities are accepted as a part of agreeing to do a particular job, whether or not we actually like or enjoy each and every one. In addition, those things that we are responsible for are those actions and events that we cause or have some causal relationship to because we are the moral agent who brings them about. This is the received view of responsibility.

The word "responsibility" operates in at least two dimensions. It contains a moral notion of personal culpability and a legal notion concerned with collective organization and the indemnification of victims. Individual moral responsibility is first concerned with imputation of blame because of an act that violates an obligation. To identify one as a perpetrator means to make him responsible for the consequences of an act. He answers for it because the action is attributed to him. Although sometimes responsibility is attributed to those who take initiative, thereby assigning them adulation, we usually speak of it in the negative sense, when someone wrongs another.

The negative usage comes from the criminal law in which we assign a penalty to someone who is guilty of an offense. We show that a person has committed an act while in control of his faculties and assign him punishment in accordance with various factors, the most important of which is the attribution of responsibility for the act and its consequences. Responsibility establishes a three-part relationship between the responsible person, the domain of his responsibility, and the authority before which and whom that person is answerable.

Legal responsibility implies an obligation to restore damage caused to others through one's own fault and in criminal law to suffer the established penalty. In morality, as well as in law, the paradigm case involves an informed intentionality, or lack of regard for the interests of others. In the professional, social, or political domain, the moral reference of responsibility for one's actions is founded on the obligation not to harm others and to restore the injuries one has caused. It is an individualistic responsibility born out of the obligation to answer for one's acts and to act in a way such that any other person could act in a similar manner. Each role one plays comes with a certain set of duties, and a person thereby assumes responsibility for faithfully discharging the duties that come with the role. This notion also includes the ideal not to charge others for one's own responsibility. It is very important to note that even if our behavior is highly constricted within the groups, communities, societies, and larger groups to which we belong, we still cannot legitimately renounce our capacity for autonomous behavior. We always have choice.

What is very interesting is how modern technology and the rise of super- corporations has put into question the traditional notion of responsibility. In this new notion the fault of the origin of an accident is not emphasized. Instead, the focus is on restoring to the victim what was taken from him. The increasing complexity of human action in modern industry, where there are often several structural, organizational, and individual contributing factors, makes it very difficult to assign blame to one specific person. The sheer complexity of our technology has, in some cases, made the identification of direct causal linkage a nearly impossible task. We see, then, in the post-modern era, a real solidarity—of a sort—in human action. We question that this appearance of solidarity is real solidarity. In this new paradigm, employers pay heavily into insurance funds, which are in charge of paying damages to the wounded. We replace individual fault by social risk management at the corporate-managerial level, but this doesn't alleviate the notion of responsibility at the individual level. This has effects in our attitudes toward non-human, sentient primates. We can also see how the growth of human population and the increasing complexity of social and political groups has also contributed to our understanding of the notion of responsibility in the same way. A consequence of this is that the human population is disempowered to respect and honor any kind of responsibility that moves beyond negative responsibility as explained. Thus, as we continue to read further into this document, we must begin to consider other understandings of our responsibility.

Solidarity
&
Responsibility

Solidarity creates a system in which one can be responsible without necessarily being guilty before the law. That is, individual responsibility persists, even though because of the complexity of the modern corporation and a growing human population it may be difficult if not impossible to assign legal blame for negligence or other malfeasance. Civil law, which insists on restoration, develops much quicker than criminal law, which is founded on penalty and on the pursuit of the perpetrator. Whereas criminal law and morality may often converge when determining the illicit act, civil law reinforces a disconnection from morality.

The notion of responsibility for risks develops a natural solidarity among the members of a particular community—the risk is shared collectively. With a large human population, this can create a tendency toward irresponsibility of individuals. This is actually a shift in consciousness through self-deception, which alienates individuals from a fuller and richer notion of their ability to respond. A new understanding of solidarity can overcome this alienation but it requires transcendence of the law, and to some extent, morality, as we seek an understanding at the ontological level. This is a matured solidarity that comes from a deeper appreciation of the whole and its natural truth of ontological parity. Because it is so difficult in some cases to assign individual culpability, we must at least consider the benefits of struggling together to deal with and overcome these complicated misfortunes, for they impact all of us and we suffer together, more or less.

In the new society, with its complicated industrial-technological complex, many sorts of human relations are no longer face-to-face. We affect others in every aspect of our lives, therefore, even though we do not see them. This applies to non-human sentient primates, too, for in most cases most of us will never see them but only hear of them generally in an oblique way. In some cases, we see them or pictures of them imprisoned in zoos, and therefore have either a superficial view of their needs as sentient beings, or we view them in distorted ways, far from the truth. This has lead to a decreased sense of responsibility amongst individuals. This leads to the danger that society itself will suffer the burden of taking care of the risks, and because of a diffused sense of responsibility, it is and will be difficult to protect the interests of non-human primates.

In addition, we must also be on guard against the transfer of responsibility to those who are perceived as being responsible, for example, government leaders, scientists, corporate heads, other professionals—anyone but us personally. This tendency here can lead to witch hunting of "those responsible," who ultimately perhaps may answer for the negligence of others. In this regard, Taine himself claimed that: "The popular imagination needs living persons to whom it can impute its wrongs and on whom it can unload its emotions." Here, restoration is viewed as insufficient, and punishment itself becomes the goal, in the attempt to purify society of the evil that threatens it. In this paradigm leaders are at risk as the body politic, including workers, members of professional organizations, business partners, students, family members, and more, each give up their own accountability. As we use the Internet and other advanced communication technologies, we face the risk that this phenomenon may become more serious and more dangerous.

Nevertheless, behind the veil of anonymity there are real and actual persons who assume various responsibilities and who act them out in situation. The ones with more power and information have responsibility for the execution of tasks that have been entrusted to them and for the wellbeing of the persons—and in our view, all primates—for whom they care. While denying a direct causal relation, which is one of the characteristics of moral imputability, this responsibility dissociates itself from civil responsibility, which we have already viewed as a collective insurance. In this way, responsibility is understood as an emergent quality of playing a certain role. Further, by diffusing responsibility, we lose a psychological imperative for taking action when we should, thereby engaging in bad faith strategies to pretend that we don't have such obligations. In street vernacular, it is always "someone else's problem."

In professional or occupational ethics, the level of responsibility of a particular person is correlative with the degree of power, knowledge, and liberty that are tied up with the part he or she plays within a certain organization and with the type of services he renders.

This often corresponds with a certain level of income and social power, as well. For persons who practice a profession—a doctor or a lawyer—moral obligations are strictly and clearly framed in terms of individual responsibility. Thus their acts are more easily understood than those of workers, whose power and behavior is always contextualized within a long chain of interaction and whose productive work cannot always be seen directly. This phenomenon deepens as we focus on the general actions of citizens in general. Again, in this system we lose a personal sense of responsibility to the whole and therefore live primarily in a complacent state.

In addition, there are ways of administering human resources that reduce the responsibility of the agents involved, insistent as they are on obedience to instructions and subordination. This does not mean that the public or that the company itself does not look for a scapegoat in its workers or citizens, even though the actual tasks that they were responsible for were not substantial enough to have caused the malfeasance. We must always look at the organizational character, too, for some have rigid pyramidal structures and others have more cooperative and participatory, which allow for more expression of individual autonomy and responsibility. This triggers the question about noetics, which can apply to individuals, to companies, to societies, and to whole cultures. Because of these new complexities, some of which we have outlined, our responsibility forces us to consider our underlying philosophical anthropology, which affects individuals as it does the whole. This exegesis is a preparatory thinking about a new noetic foundation for the self, intersubjectivity, and the relations of humans to other life forms, specifically focusing here on all primates.

In a complex system in which everything is more entangled and more uncertain, it is important to acknowledge and promote the autonomy of those involved. This makes apparent the need for clarification of the role, power, and autonomy of each position in the web of relationships, which in turn creates a real ability to assign responsibility. Concomitant with this structural organization in which autonomies are created, we need a clear statement of standards of conduct, for each role to be discharged in the system. This set of ethical imperatives becomes a promise that each of us makes to each other, by playing the role that we agree to and to which we are assigned. Clearly, these promises need to be re-evaluated periodically, and we also must provide the sort of training and education that will enhance one's ability to carry out his or her duties. In short, responsibilities must be clarified and the network of all interrelated responsibilities communicated in such a way that we all understand how our duties fit in with the whole. This imperative overcomes complacency.

Moreover, we must always be aware of changes in global conditions in all domains, including the economic, the governmental, the social, the legal, and the scientific, which may transform relations of relative power in such a way as to create new zones of responsibility. These new zones of responsibility alter current moral relations as well as create new ones, thus we ought to always be vigilant. Furthermore, we cannot ignore those who are distant from us either in space or in time—or who are not humans. Just because we do not ever see someone or understand his or her life system, we may have moral responsibility to them because of the complex interrelationship of all things. That is, we may not be able to see all the consequences of our behavior even though we know that they exist. We ought to also be careful about the future, for each action we engage in now has important effects for future generations. Hans Jonas expresses these concepts in this way: "Act so that the effects of your action are compatible with the permanence of genuine human life," or "Act so that the effects of your action are not destructive of the future possibility of such life." This implies that we must act on right principle because we cannot see or understand the wide-ranging consequences of our behavior. We must, however, extend this thinking to accommodate a trans-human worldview. This requires us, as we have stated, to pursue a deeper understanding of all aspects of our ecosystems, the whole of these aspects, and our place within this whole. This implies that no other sentient being, most especially other primates, may be used, disregarded, or harmed for the perceived benefit of humans. To not regard the full interest of all non-human primates means to ignore an understanding of the whole. This is an obscurantist and neglectful view that is dangerous not only to non-human primates but to human primates because it lacks, in principle, critical thinking and a comprehensive view of the interests of each group and its value to the whole. In sum, this is a form of species narcissism, analogous to the same thinking within the human population, in which some groups are wrongfully treated with greater ontological value.

This new notion of responsibility orients itself toward

the trajectory of our behavior and comports with the forgotten etymology of the word "responsible," which is called sponsor from the Latin. A sponsor vouches for someone else. To be responsible means to stand as a surety for another. It means to be capable of making promises, giving one's word, making a decision in which one commits oneself to the future. From this promise emerges behavior and one's contribution toward the situation of others, including non-human primates. Seen in this light, responsibility means that one is able to keep one's promise. Further, this is both an explicit and implied promise that serves the clear trans-human values delineated in the preamble. Concomitantly, it means to inspire confidence in others and to encourage them to pursue their own virtue. Most importantly, it is about a lifetime exercise in virtue, in which we do work on ourselves, in relation to the others in our lives, and the radical Others we will never see, hear, or know. At its very roots, then, freedom is mediated by responsibility to others because we are always in relation to them. In short, we must overcome our species narcissism and the possessive-self doctrine of the Western Enlightenment.

Let's take stock of our account of responsibility. We must, first, always carefully consider the effects of our behavior on others, present and future, near and far, familial, familiar, and exogenous. We must also take steps to prevent any foreseeable harm, before we act. Not to be aware of danger is irresponsible and thus we must live a life of preventative prudence. This obligation extends to all the domains of social life even where the causal links are harder to see. Second, we are witnessing an explosion of knowledge about our world, and this also mandates that we engage in reasonable research to discover potential harms. We must derivatively pay attention to the obligation to communicate both our knowledge and our lack of knowledge about any factors that may affect a risk of harm. The third obligation consists in restoring to victims what they lose, through the system of suretyship. Presumably this transfer of resources back to victims comes through channels of compensatory and even retributory justice to which we adhere and about which we dialogue, keeping firmly in mind that education and moral growth is the main objective. This means that we must consider both civil and criminal penalties for any harm to non-human primates in the same way that we consider them for human primates. Justice requires it.

A New View of Responsibility

Traditionally, ethical responsibility focused on antecedent causes and was mostly concerned with individual duties that emerged out of present and past relationships. This paradigm is giving way to a new focus toward the future and long-term relationships, especially pinpointing those individuals who are especially susceptible to the effects of collective action. We are moving from a mere system that prohibits us from actively harming Others to a system in which we are responsible for proactively changing structures that harm and for actually helping Others, including non-human primate Others. We are, then, not just responsible for keeping our house in order. The watermark is higher, and now we are more and more responsible for Others—those we supervise, the innocent, the weak, those in the future, and again, our non-human primate relatives. My responsibility no longer emanates from inside of me. Instead, it comes from each other primate—both human and non-human—who calls to me, who appeals to me. I, in turn, must listen attentively so that I may hear their calls. I must look beyond my own immediate circle of relations and adopt a broader view, becoming responsible for a much wider range of individuals. This includes all sentient Others. This includes all primates. It is not that we become responsible for everyone and everything. Yet, it means that we transform a duty-based ethics where we pursue the right action into the pursuit of virtue in which we pursue the good and are always on the look out for the needs and interests of others. This includes all primates.

Let us consider the developmental process of assuming broader and deeper levels of responsibility in a mature way. Before we are conscious, we are beings with urges and wants. This is the state of a child whose original urge is for self-directed happiness. We can characterize this wanting as selfish, narcissistic, with the logic of pursuing as much pleasure as possible and as avoiding as much pain as is possible. Here, the conscious subject thinks that he is the center of the world and the measure of all things. He is the center of all meaning and all claims to truth. This, we hold, is the unconscious narcissism that has determined the entire western vision of who we are. It is the basis for our anthropology of the subject, of the self, and of the way we have heretofore regarded non-human primates.

In this state, we are so involved with ourselves that we view all other things in light of ourselves. We are shackled to our own self-image and find our wellbeing in the pleasure we have in our own image. At bottom our desires are directed right back toward ourselves as we each try to understand the world in terms of our own mental categories, including other people, other sentient beings, and non-human primates. In this state, then, we totalize a world in terms of our own small frame of mind, our narcissism. Here there is a complete convergence between oneself and objective reality. This lack of a gap or discrepancy is our propensity toward self-deception. It is a denial of the transcendent, which is the world of possibility, of overcoming, and of change. This foundation creates complacency, depression, and misplaced aggression. It also creates alienated, incomplete, and distorted consciousness, which leads to false beliefs about the nature of the world and the value of other sentient beings, including non-human primates.

This state prohibits us from directing ourselves toward another person, and toward a member of a different species with respect for his difference. Rather, in this state we utilize the Other's image for our own project of becoming. In this kind of way of life, which is mostly unconscious, I want Others to want me in various ways. This helps me develop my self-identity as well as my self-esteem. Yet if I approach Others in this narcissistic way, only in terms of my own need for wholeness, I preclude myself from listening to their need. This way in the world creates a kind of tit-for-tat, competitive [and adversarial] social dynamic in which we do things for Others in the hope and with the expectation that we will receive something in return. It creates a world that is a war of all against all, overtly or covertly. For beings we regard with less ontological value they ineluctably suffer from our narcissism and our deep needs to create ontological wholeness by harming them.

Unfortunately, when we see Others in this way, we often unconsciously attribute to them the qualities of ourselves. This externalizing of our own self creates a project of mastering the Other when in reality we are just trying to master the wounds of the past that we have internalized into ourselves. This kind of need can also show up in the belief in a god that is narcissistic. In this state, we pursue a unity with god in order to obtain an overwhelming assurance and protected-ness. Again, though, in our narcissism we are only chasing ourselves. It remains to be seen and understood whether and how our harm toward other primates emerges from our own need for self-identity, power, and protection.

What becomes a roadblock for these ontologizing projects is our realization that we can never reduce the Other as a function of our own happiness. He is always different, and always escapes what we want from him. In the case of our treatment of non-human primates, we are just beginning to understand our difference and our similarity, more importantly how we have used and harmed them in the service of our own hubris and narcissism. It is also true that we have not gotten what we want from them. This is why we are relentless in their exploitation. More deeply, we hold at the very same time that they are like us and yet not like us. We say that we can gain valuable information from doing harmful and noxious experiments on them, and at the same time that we are not being immoral because they are different. We certainly would never perform these sorts of harmful experiments on humans; moreover, we would never imprison humans the way we do non-human primates, disregarding their basic interests and needs that are no different than ours.

Yet, in this egocentric involvement with the world, we create relations of power with other people, more so with non-human primates. Because of the fear of losing our own identity in the face of the Other, we try harder to ignore our difference, but in this way we don't see what he may need from us. This creates a social and interspecies world that is very much based on an economy of usefulness, encouraging us to see the Other in terms of his or her use value. I use you in order to affirm and develop myself; you do the same; and so do we all. We see this within human primate communities; we see it even more in human treatment of non-human primates.

All of us, led by our egocentrism, are motivated by our need for autonomy, wanting to expand the reach of our influence as much as possible, and thus we collide with each Other in our everyday push for power. From this conflict the numerous people who inhabit the same ecosphere cannot each be at the center of things. This causes us—in our egocentricity—to position ourselves against each other. *A fortiori*, it causes us to treat non-human beings as things that can be exploited for our perceived benefit. This extends to other primate groups. But hardly anyone wants to live in this kind of system, in which no one can be trusted and where we constantly have to watch our back. The better dimensions in life—peace, music, relaxation, pleasure, and the like—don't come when we are always on the verge of violence. As a consequence, we enter into compacts with others, giving up only so much of our autonomy as is necessary to avoid the other's violence. This leads to the formulation of laws and clear statements of rights. So far, we have only started to protect all human primates from predation and use-abuse. To date, we humans have been very begrudging about extending this protection to non-human primates.

The problem with treating radical Others as things to use and exploit is that it treats weaker groups, weaker individuals, and other species groups and members unfairly. This is so because the system itself is born of egocentrism and relations of power. Within the logic of this system, attention is given to the powerless only to the extent that they pose a threat to the powerful. Our history, then, is a history of conquests; it is a humanism of those that are proud of being conquerors. Within a history of this kind of competitive dynamic—of all against all—we have become desensitized to the Other, and the cycle intensifies. Most of us suffer a constantly nagging that something is wrong, but we hide from it; we run from it. We use food, drugs, sex, shopping, and the like to distract ourselves from the truth of our lack of attention to the Other. We thus retreat from our responsibility to Others, each Other, and all Others. Until we change the basis of our thinking, we will cripple our ability to see the violation of the natural rights of other sentient life forms, including non-human primates.

The Look of the Other & Maturation

Nevertheless, at some point in our development, in our maturation process, we understand that we are not at the center of the social world, the ecosystem, and the planet, and that we never were. This presents us with the conceptual and experiential problem of finding a way out of our narcissism and our egocentrism. We cannot do it in terms of our own internalized, ontologically-deficient, pleasure-seeking enterprise. It therefore must come from outside each one of us. It must come from the Other himself. As such, we must transcend viewing the Other as a force that must be overcome or that we can own and possess to his or her detriment. This applies to all primates. Furthermore, we must see the Other—not in terms of our own categories of understanding and not in terms of our own agenda—but in terms of his or her radical difference.

This transformative understanding comes from the very look of the Other. It is the other's glance that breaks through my egocentric construction of him. When I look into the Other's face, directly in the eyes, I see how different he is. I can flee from this fact. I can hide it. But I know it. This is the fundamental fact in my interaction with the Other. I know at some level that he always escapes me, always escapes my ideas about him, and that he lives in another universe of being. It is an uncanny feeling that we experience, as we realize that the Other really is different. Just the look of the Other resists my attempt to wrap my head around him. It is his plea for me not to annihilate him—not to kill him. That he is unfamiliar. We see this constantly when we look into the eyes of the other sentient beings we exploit. We see this in the eyes of all primates.

The face of the Other becomes my judge. He raises the issue of my power. He raises the issue of his own difference. In the Other's face we recognize that what we take to be our right and our freedom is just an imperialism that conquers, that takes, and that murders. I realize that my spontaneous impulse toward self-affirmation and self-development is not so innocent, that I affect Others from the ground up, even in my so-called reality as a free being. My conscience tells me from the bottom up that I am necessarily interdependent with Others. This includes all sentient life, including humans and other primate beings both known and unknown.

This responsibility that causes me shame precedes my ideas about freedom. It precedes any agreements I make. It has little to do with feelings of altruism or sympathy for Others, feelings that come from conscious awareness. I learn that my responsibility comes as a very part of who I am. I learn that I am not independent from Others and that I live for the Other—one way or the other. I learn that I am not only responsible for my own "free" decisions but that I am also responsible for the Other, and for his responsibility as well. Dostoevsky himself agrees with this extreme view of responsibility by saying in The Brothers Karamazov that "each of us is guilty before all, for all, and I more than all others." I am thereby responsible for my neighbor. When the Other looks at me, calls to me,

I cannot avoid him. I cannot pretend that responsibility for him is not my concern for my pretending belies my acknowledgment, that I have heard his appeal. In the midst of the Other's face I realize that I do have a choice. I can either use him economically or answer to his appeal. If I answer his appeal, I act without regard for my desire. This goodness demands no repayment or satisfaction. Here, the concern for myself disappears from the center of attention. I become preoccupied with the Other's situation and the Other's need. I surpass my concern for my own life as I concentrate on what the Other needs from me. This responsibility extends to all humans, including the elderly, the weak, the infirm, and the disabled. The law that we here promulgate further extends this responsibility to non-human primates. They each have a Look, and they each appeal no matter what their level of functioning, consciousness, ability, or species status. They each play a part in the whole of primates; the whole of the ecosystem; the whole of the planet. Moreover, there is no good reason to preclude non-human primates from these protections. If we extend the map of interdependency beyond human beings in order to include the biosphere as a whole we see that the interdependency goes beyond humanity, returns to it, and is informed by it. Our humanity is thereby deepened, clarified, and improved.

Let us be clear that this struggle involves the choice between egocentrism and responsibility. Goodness comes every time I successfully free myself from my selfishness. I must constantly be vigilant to my natural tendency to put myself first. The Other must always count for more. Perhaps goodness just is this struggle between narcissism and responsibility. This is not to say that I negate myself as a person or that I negate my own identity. Instead, it is a matter of putting the Other first as a matter of course and as a reversal of spontaneous, selfish inclination. Giving to the Other goes hand in hand with emptying myself of myself. Giving truly means giving something up. Giving means giving a gift of myself. When I am responsible I am for the Other always, in spite of myself.

In this new paradigm, or way of being, I do not primarily seek out the protection of my own rights. Instead I seek out and advance the rights of Others. I protect the Other first. I serve the other first. I look out for the Other first. I answer to the Other first. I give to the Other first. I give to the Other without regard for my feeling. I give to the Other without expectation of repayment. I give to the Other without considering the potential advantages. In my giving, which is without regard for any instrumental good that I could achieve through my giving, I give universally and without discrimination. I give because it is my responsibility to do so. Thus, my relationship with the person or sentient being to whom I give is not a relation to a particular Other; instead it is a relation to all, distant and near. It is a universal stance.

When I am responsible to the Other I am led by the truth that the Other has a right over me. All Others occupy this place in me, thus I must act in ways that protect even the absent Others. This responsibility occurs and animates itself through any dimension in which I can protect the Other, which includes the social, political moral, scientific, professional, practical, and others. This requires careful mediation between all my responsibilities to Others, as well as to my responsibilities for my own self-development. I must develop my abilities to serve Others. Each of us, to whatever degree we are capable of must also, and especially if this is the role we are playing, make sure that the institutional structures that carry out justice are based on the kind of responsibility about which we have been talking. One can argue that this kind of a society is a utopian ideal but this is not a good reason for not trying to attain it. We must constantly charge ourselves to higher standards, as long as they keep us from drifting back into the kind of egocentrism that leads to the sort of violence and war that we see today.

Solidarity and Rights of Both Human Primates and Non-Human Primates

Now that we have established an account of solidarity, we can explore how it can be understood as the basis and foundation for both human rights and animal rights. As has been shown, through phenomenological investigation, we see that there is a basis in human consciousness and human anthropology for the value and virtue of solidarity. Justice is the recognition of solidarity, human rights, and the rights of all primates.

In the *Preamble of the Universal Declaration of Human Rights of 1948*, it is said that "Whereas recognition of the inherent dignity and of the equal and inalienable rights of all members of the human family is the foundation of freedom, justice and peace in the world." The framers of this prolegomena believe this foundation must include the equal and inalienable rights of all primates in order to lead to a fuller concept of freedom, justice, and peace. Utilizing the language of rights is particularly powerful because it best helps us understand the relation between political structures and individual moral agency.

In order to focus on rights, we must first look to individuals, for it is the human person who is at the foundation of the dialectic between autonomy and heteronomy, solidarity, and rights. It is the individual human person, as moral agent, who is the most basic unit of society even though we all live within complex structures, both conscious and unconscious. We think that there is a close connection, both existential and rational, between the conversion to heteronomy, a commitment to solidarity and responsibility, and the furtherance of rights-based understandings of justice that promote the interests of all primates. Solidarity lies at the foundation of freedom, equality, and justice, and implies the values of mutuality and reciprocity. This means that humans must work toward a world that is sensitive to and supports programs of mutuality and reciprocity toward all primate groups. This is a relational view of our moral imperative, which we have considered in our earlier charge that we must gain deeper understanding of the whole and we must allow the Other to interrupt our being.

In a trans-human perspective, there is an argument that a fuller understanding and experience of the human person involves the conversion to heteronomy, which includes a full commitment to solidarity and justice. This avoids the dangers and lack of a system of justice and rights based on the sort of autonomy we contemplate in a humanism of the "I." This leads to a greater equality and freedom for each and every human and non-human primate, and starts with solidarity. Let us first address human rights. They are *thought* to be universal, applying to everyone no matter their country or place. They are thought to have their origins naturally, from the very fact of being an individual human being who participates in the world economy of peoples, they imply reciprocal duties, i.e., a set of responsibilities that is correlative. In this system, the virtues of reciprocity and mutuality are essential, implying an "ontological parity" or "ontological equality." There is, therefore, a close connection between reciprocity and dyads, and an extension of reciprocity and mutuality to all individuals in the human community. Thus, in each of my interactions I must treat the Other as if I am treating all Others. This makes my behavior universalizable. There is, therefore, a close relationship between an individual's striving for the human rights of all, just as she exercises them for herself; this guarantees the good of the community.

This work toward strengthened human rights requires that we develop our own and those of others in a way that fosters more participation in the community. This respects both the common good and the wellbeing of individuals within it. Here we can see that respect for human freedom and choice goes beyond an admonition to prevent it. Instead it requires that we constantly engage in behavior that develops these freedoms in ourselves and in Others. Implicated in this requirement is that there is something good about the pursuit of human flourishing both in individuals and in the community. Given the connection between solidarity and human rights, such connection emerging from a certain anthropological conception of what a human being is, what flourishing is, and what community good is, we can see how the virtues of reciprocity and mutuality, along with the duties that I am articulating, lead to more "correct" conceptions of this anthropology. This leads to the notion that there is a space between what we are now and what we could be, given an idea of individual and cultural development. When we frame our duties positively—and specify them as solidarity, mutuality, and reciprocity—we see that we have responsibilities toward the development of individuals and communities, and positive requirements to enhance their situations. This becomes an aspect of all rights. We now see how important it is to frame an expanded notion of the human being, for it is this conception that determines the extent of our responsibility. Our conceptions of dignity come from our anthropology of the human. Our conceptions of the moral, whether in our personal lives or in our professional practices, are foundationally determined by our interpretation of the human: our anthropology. Let us proceed further to see how a deeper understanding of our anthropology and what is required for human rights can also be extended to all primate groups.

Natural rights are those rights that come from our conception of the human being—those essential qualities toward which we aspire, and which we articulate through discourses of freedom. Human rights, most often manifesting in laws are specifications of those natural rights, including participation, education, health, justice, and the like. Underlying these structures of rights, however, is the complex relation between solidarity and our anthropology of the human person. Solidarity requires an apprehension about what is important to human beings, what is essential to what we are, what constitutes their flourishing, and what leads to its development. At any point in time, a human being will have reached a certain level of development, participating in the potentials of the good to a degree. Yet, there is always a deficit, a lack of wholeness in that individual. This is the area for self-development, and correspondingly, an opportunity for others to exercise responsibility toward its advancement. This is the realm of our dehiscence and our possibilities toward a kind of self-development that promotes others. Thus, solidarity promotes the good not only of individuals but also of the communities in which they participate. The right kind of community enhances the freedom and the development of all the individuals within it. This implies a pervasive interdependence between all individuals and between individuals and the community. The ultimate goal is that each and every individual becomes more human, which is the cynosure of this prolegomenon for a reconstituted (trans-) humanism that protects all primates.

Within this line of thinking, solidarity becomes the goal of communities, promoting the highest levels of self-actualization and participation of each member. It is this anthropological conception that we need to discern. The common good becomes more important than anything else because it leads to the freedom of members of these communities. Solidarity deepens and broadens those pathways and possibilities for self-actualization. This is what we mean by freedom. Importantly for these arguments is that no one is excluded, even those who cannot exercise their right to self-determination and even those without rationality, a different rationality, or lower levels of rationality. This implies that an essential element of the self—of our humanity—is the participation in communities that promote and facilitate self-development. This creates a prescriptive model of action that promotes the interests of communities and individuals concomitantly. Furthermore, it means that my humanity is never mine in isolation—it always involves that of Others. As such, I must always to promote the common good, the good of other individuals, and to oppose structures and behaviors that do not promote these goods. The common good must, therefore always outweigh individual good. Only in this way can we build communities that promote a fuller and more complete concept of the human, which is the foundation for thinking about global ethics. In terms of this prolegomenon, the trans-human perspective includes other species in a broadened and deepened conception of solidarity wherein our conception of our own humanity changes to include the interests of other primates. In short, humanity does not exist in isolation, but always in relation to other species, especially other primate groups.

This conception of a radical interdependence is not the same as a free market that promotes hyper-individualism because in individualism there is lacking a notion that community members have a common destiny. It is the opposite with solidarity and interdependence. Furthermore, any structure less than a full interdependence lacks a full and complete notion of what a human being is. For example, it is difficult to argue that a human being who ignores the plight of others or who does not desist from taking too great a share of the common good exemplifies that anthropological conception we are trying to discern. We can see together how easy it is for people to be swayed by compulsions and addictions that do not promote the common good. Quests for power, money, property, and pleasure that detour from interdependence deny a common destiny, and violate the spirit and reason of solidarity. Without solidarity most if not all humans will not be able to self-actualize, i.e. to become authentically human. Thus, living out all the parameters of solidarity is just the highest level of self-actualization, which at the same time, promotes that of all Others, and serves the common good as community. For all the reasons stated, we believe that these ideals of individual and social actualization must now include all primates. A more complete anthropological conception of the human demands it.

Solidarity is perhaps a richer concept than first meets the eye because it requires a richer type of optic experience. It is more than a reorganization of wealth, opportunities, or goods. It is a way of being, and determines how we see ourselves, the world, and Others, in a way that goes beyond an instrumentalism. If we view each of us as equal, deserving of the same treatment, substitutable for each other, and holding ontological parity, then we can begin to understand our radical interdependence by choosing for each of us, ourselves, and all of us simultaneously. We come to see others—**all primates**—as both objects and as subjects, each with his or her own life plans, goals, and preferences—ones that may not agree or comport with our own directions and projects. Foundationally, however, we must identify all of the important common elements to human **and all primate** being-ness, such as agency, safety, water, food, shelter, sociality, and so forth; it is to these needs in all of us that we must focus our attention, orientation, attitudes, responsibilities, and duties.

What protects the dignity of **all primate** individuals and communities is, therefore, solidarity. It coaxes us, and even demands, that we assume increasing levels of responsibility to the good of Others. This requires the awareness that we are always in relation to others, including **radical Others**, and in relation to our communities. It includes the assumption that we are ontologically equal, all of us human primates and non-human primates, and with all other sentient beings. Finally, it requires us to learn about and then practice a certain group of enabling virtues, such as compassion, kindness, mutuality, reciprocity, and empathy, in order to understand human and non-human primate dignity and the responsibilities it imposes.

Now that we have established an account of solidarity, we must see how it leads to our quest for a new human anthropology, one that develops from autonomy to heteronomy, and then beyond. The meaning of being a human being is elastic, and can change depending upon a number of factors, including culture, geography, ethnicity, personal history, and the sort of anthropological thinking of this prolegomena. We know that there are structural factors about our human being-ness that we cannot avoid; however, there is an element of choice in what we want to be; into what we wish to change. This includes the way we value other life forms, including non-human primates. Because there is an inherent human faculty to strive toward various ideals and goals, we think it academic that humans can strive toward new and better conceptions of their humanity; their human-ness. The challenge is to identify and define what the essence of a human is. So far, we have been following a line of reasoning about what humans are that follows a very different worldview than those who believe that essence of humans is to be competitive, adversarial, and violence. Upon reflection, however, we have seen that this negative view of humans is a distorted and under-developed view. Through reflection we can see that all of the characteristics associated with solidarity are essential qualities to a more accurate and comprehensive understanding of human anthropology. This requires that we come to know what is authentically human and what is not. In our prolegomenon, we argue that authentic human beingness requires us to value other, non-human primates with ontological parity.

Let us now focus the discussion on our relationality with Others, which includes all other human primates and all non-human primates. This is our sociality—our socially embedded nature. If it does not happen in the right way, then we cannot be individually authentic, so this is a major structural element in our understanding of the relation between rights and solidarity, and gives us clues about our anthropology. This means that our individual connections to equality, freedom, solidarity, responsibility, and rights are always contextualized in terms of that sociality. This can be a starting point for delineating an account of authentic human experience, i.e., our anthropology. There are, of course, other dimensions such as spirituality and uncertainty, the material world, self-introspection, mortality, temporality, cultural requirements, and so many more. Nevertheless, for the purpose of this prolegomenon and its reflections, we must focus on sociality and interdependence, for we can only clarify our personal identities by understanding how they are located within our embedded social, cultural, and historical nature. It is these underlying historical and cultural structures that determine the range of possibilities for our freedom and therefore the answers to questions about who we are. Furthermore, it is within these ranges of freedom that we can locate pathways toward the Good, for ourselves, Others, and communities—**including both human and non-human primates.**

These social structures of embeddedness precondition meaning structures, ideologies, frameworks for freedom, and a large number of personal identities that condition moral agency. Striving toward a more fully human experience means that there are more opportunities for self-determination in the communities in which one lives. We can plainly see here that a person can only be self determining if he or she knows his self. This is the quest for a more authentic experience of life—and of one's self. Yet, here we face a problem. This is the epistemological problem of how one knows her self. Kant himself constructed this problematic in terms of rationality and the promulgation of duties toward other rational agents. This is how he created his notion of a community of ends. However, in so doing, he elevates the faculty of rationality to a godlike status that becomes a mixed bag of consequences in the way we think about our human being-ness. Unfortunately, this elevation of "rationality," whatever that is, subjugates other qualities, humans who are less rational, and all other sentient life, including non-human primates—beings who/that by definition are less. Thus, we have a distorted view about dignity and cultural goals that do not always include respect for human and non-human primate dignity. Instead, we have a system that pays homage to dignity even though its economy does not stem from ontological parity. It is a subtle error, but one that gets covered up in Kant's maxims. To be sure, we ought to encourage the kind of rationality, i.e., a kind of critical thinking that promotes the Good. Yet all too often what passes for rationality is relational power in its various guises. Moreover, humans who have less of this faculty are subtly subjugated. Sentient beings who are not humans have it even worse: because they are not humans it has been thought that they don't deserve any inherent dignity, other than derivative forms of sympathy that might lead to thinking and behavior more in line with universal maxims.

On the positive side of this, to the extent that we are rational, we can create a mutual and reciprocal system of rights and duties that protect the dignity of all human primates and non-human primates. The pursuit of solidarity then becomes an essential element to the construction of this system and our responsibilities to it. Because there is such an interdependence between all primate beings, a fuller anthropology must not only specify our negative duties and responsibilities, i.e., our duty not kill, but it must include strong duties and responsibilities to promote the development of Others. This requires human primates to advance the interests of non-human primates. Typically, we view our freedom as unlimited to the extent that we are not harming the interests of Others. However, in this new anthropology we have much stronger duties to promote the interests of Others. This includes a stronger sensitivity to our relational character, which we recognize as essential for our fuller development as human beings and human communities. A simple way of re-articulating this point is that we simply cannot achieve fuller and more complete levels of our human experience except as we take on deeper levels of responsibility to Others, to our communities, and to all primates.

Because we have been sedimented within a certain conception of what a human being is, and what human communities are, we have not been able to see these other ways of existing within a richer anthropology. As such, we have not understood our fuller responsibility to solidarity, or how to live it. In order to do this, we must disengage from the kind of subjectivities that we have created, i.e., replace the phenomenological filter on our conception of the human being, and replace it with a new conception. In short, we must let go of our attachment to the possessive, egocentric self who lives in a radically individualistic way. We must substitute it with a self that understands our relational nature, the essential recognition of our interdependence, and therefore, one that develops a strong motivation to live in accordance with these truths. Currently, we live in a mostly zero-sum, highly competitive world that is structuralized in many ways against the kind of vision of culture, society, and self-hood that trans-humanism attempts to articulate. It produces alienated, isolated, highly individualistic beings instead of those who recognize their need for each Other within a highly dynamic sociality, interdependence, and solidarity. We maintain humbly that as a whole we simply don't see the possibility of this new kind of social life and human anthropology. We are therefore blind to our own self-potentials.

In summary, our quest is to discover what is most human in human beings. We see that the valorization of autonomous individualism does not reveal what is most human. Therefore, we attempt to seek it in heteronomy, which triggers our thoughts about solidarity, rights, and interdependence. We understand that we are co-constituted with Others and that our autonomy and freedom never arise without this co-constitution. We can also see how solidarity is foundational to the actualization of both freedom and equality. This to understand how the fate of one is tied to the fate of the Other and that through dialogue and participation we can and should pursue the good of the Other as we pursue the good of ourselves. This means that we re-interpret our responsibilities in terms of a much more pro-active sense of solidarity-in-action. This goes back to our earlier account of desire with the modification that we take into account—first—the interest of the Other. This is heteronomy. In terms of individual subjectivity, we develop a dialogue within us that participates with the Other and with the community—internally. In this way, we become responsible for all Others—all primates. Thus, while we pursue our own values at the very same time—because of that internal dialogue—we actively seek ways to promote and facilitate the interests of Others. This leads us, nevertheless, to a new conceptual knot, one that we must unravel in order to arrive at this new anthropology of which we speak. This leads to a new double problem: We must still determine how to reconcile the interests of self and the Other, i.e., resolve the dialectic of autonomy and heteronomy. We must also figure out what to include in the ambit of "Other." In this prolegomena, we consider the interests of all primates both human and non-human. This is an approach to human subjectivity that eschews the autonomous subject of the Enlightenment in favor of a vivantonomy that recognizes the ontological parity of all primates.

The approach this prolegomenon takes is to examine

the necessity of a change in human subjectivity—our anthropology. Now that we have seen the improvement a heteronomous anthropology has over autonomy, we must now explore and identify how to use these insights to reconstitute the autonomous subjectivity we have created. We do this through a trans-humanist approach that moves beyond the direct interests of human beings and that includes all primate life, equally. This is trans-humanism, a theoretical foundation that adopts the principle of ontological parity; a new formulation of responsibility and solidarity; a new commitment to knowledge of the whole of our ecosystem; and a new Archimedean point for our role in the universe.

By understanding a new ontological approach in our relation to non-human primates, we allow our own narcissistic project to be interrupted, thereby re-envisioning new understanding of the value of the Other. In the phenomenological encounter with the radical Other—the non-human primate—we are led to a deeper level of responsibility and solidarity. This is the basis for this prolegomenon, which exposes the truth that there is no salient, real, or relevant difference between human and non-human primates. This implies a phenomenology of reverence.

Conclusion & Implementation

We recognize that this prolegomenon and proto-law with philosophical exegesis and argument must be considered with great care and reflection by thinking people. In a historical sense, the time is right and appropriate for the law of this land to now articulate these principles toward which we have been driving for many years. The goal is that the great politicians of our time will see the wisdom in our approach and the simple pragmatic truth we espouse. Like other beings who have been treated unjustly, non-human primates have suffered long enough, and their natural rights must be propounded by a federal law that unequivocally states it as such. Furthermore, it will take much discussion to fairly implement new law and to allow for further refinement in the years that pass. However, this is not a reason to not begin this process now. Like the wise anti-discrimination laws that we have passed recently in our history, this new law follows their logic and ratio, and must be considered timely and urgent. It is the framer's intent to prepare a second edition of this prolegomena with historical, legal, moral, and philosophical annotations; definitions; and concrete examples. This is the next step toward righting a great wrong: a moral catastrophe.

Kevin Boileau, PhD, and Nazarita Goldhammer,
Directors of Eidetic Science & Humanism at the
Existential Psychoanalytic Institute & Society
Writing in Missoula, Montana, USA, Autumn, 2015

Freedom4Animals is a contemporary animal rights organization based in North America. Its overarching mission is to engage human beings in discussion about the moral, legal, political, social, economic, cultural, psychoanalytic, scientific, and educational aspects of how we treat non-human animals. Although F4A has relationships with other animal welfare and animal rights organizations, its members and supporters are independent in their judgments, projects, and initiatives. By working with scientists, professionals of all kinds, children, teachers, and the general public, F4A hopes to sensitize more humans about the plight that animals of all kinds face on the planet today, especially with a growing human population and therefore a great demand for food, money, and other resources. By propelling this discussion to center-stage, our hope is to elevate the level of thinking and reflection about how we treat non-humans animals and, correlatively, how we treat fellow humans.

Dr. Kevin Boileau
415.830.0065
Nazarita Goldhammer
408.499.5062

email us directly at: animalsRfree@gmail.com

About EPIS Press

We established EPIS Press in the objective of publishing new work in the following areas of inquiry:

1) existential psychoanalysis & phenomenology;

2) traditional & contemporary psychoanalysis theoretical and clinical;

3) critical philosophy as it pertains to psychoanalysis, culture, phenomenology, and philosophy of mind;

4) new literature in phenomenology and psychoanalysis;

5) any related work as it bears on these issues, including neuropsychology and psychoanalysis.

For more information go to
EPIS Press
2026 S 9th W
Suite 4
Missoula, MT 59801
epispublishing1@gmail.com
www.episworldwide.com

also by
Kevin Boileau

Theory

Genuine Reciprocity
and Group Authenticity:
(The Social Ontology of
Sartre & Foucault)
First Edition Available

The Algebra of History
(with David A. Boileau)

Essays on Phenomenology
and the Self

Manifesto on Solidarity

The Myth of Mediation
Neutrality

Game Theory for Mediation

Coming Soon:

Phenomenology &
Mediation

Workbook for
Psychoanalysis & Mediation

The Phenomenology of
Power & Violence

Literary

The Patient

The Blue Pearl

Abject Poverty

99 Deceptions

The Separation

The Return

The Outlaw Series Vol. 1

Make Me Stay

Sexxual Lies

A Turtle's Murder

Coming Soon:

KillBox

Northside

The Bishop... A Fisherman

3 Rivers

Sweet Monogamy

Made in the USA
Monee, IL
04 November 2026

67378091R00095